W ma you YOU?

ILLUSTRATED BY Marc Mones

A & C BLACK

For Tom, Ellen and Robert,
who make me *me!*

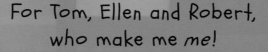

First published 2013 by
A & C Black, an imprint of Bloomsbury Publishing Plc
50 Bedford Square, London, WC1B 3DP

www.bloomsbury.com

Bloomsbury is a registered trademark of Bloomsbury Publishing Plc

Additional picture acknowledgements:
Additional images all Shutterstock, aside from the following: p6 top right and p58 left © Wikimedia,
p6 bottom left and p58 top right © Wikimedia, p11 bottom right and p58 left © Wikimedia, p16 top left
and p58 right © BE031947/Corbis, p16 top middle and p59 left © King's College London, p16 top right and
p58 right © BE031947/Corbis, p17 top right and p58 bottom © Jenifer Glynn, p19 top left © Science Source,
p19 top left © Science Source, p29 bottom right © Wikimedia.

ISBN 978-1-4081-9406-5

A CIP catalogue for this book is available from the British Library.

All papers used by Bloomsbury Publishing are natural, recyclable products
made from wood grown in well-managed forests. The manufacturing processes
conform to the environmental regulations of the country of origin

Printed in India by Replika Press Pvt. Ltd.

10 9 8 7 6 5 4 3

Contents

Introduction

DNA ... clones ... stem cells ... genetic engineering ... They're all in the news, but what on earth are they? What's fact and what's fiction; what can we do now, and what might we be able to do in the future? Growing new hearts, using sheep to make medicines, cloning pets... It all sounds like science fiction. But it's not fiction, it's fact, and it's *amazing*.

This book will take you from Charles Darwin and Evolution all the way to some of the most exciting developments in biology taking place *now*. This is a story about every one of your cells, and about everything that's ever been alive. This is the story of what makes you *YOU*. Come on into the book, and let me tell you a life story. In fact...

Let me tell you the story of life itself!

Charles Darwin and Evolution

Charles Darwin (1809-1882) was interested in all sorts of living things. He was fond of worms and pigeons, but he really *loved* beetles. When he was young he was out looking for beetles one day and spotted one he didn't have, so he picked it up. Then he found *another* new one, so he picked that up in the other hand. Then he found yet another new one, but he'd run out of hands. So what did he do? He put one in his mouth, of course, so he could pick up the third one.

Charles Darwin developed the *Theory of Evolution by Natural Selection.*

Here's how it works:

Some sorts of animals and plants only have a few babies. Others have lots. Some have *millions*. So why hasn't the world filled up with rabbits or rats?

Darwin spent his whole life studying all sorts of living things, and made many discoveries, but the one he is remembered for is called the *Theory of Evolution by Natural Selection*. It explains how all the living things in the world today developed over a huge length of time from a few, much simpler living things.

ON
THE ORIGIN OF SPECIES
BY MEANS OF NATURAL SELECTION,
OR THE
PRESERVATION OF FAVOURED RACES IN THE STRUGGLE
FOR LIFE.

BY CHARLES DARWIN, M.A.,
FELLOW OF THE ROYAL, GEOLOGICAL, LINNEAN, ETC., SOCIETIES;
AUTHOR OF 'JOURNAL OF RESEARCHES DURING H. M. S. BEAGLE'S VOYAGE
ROUND THE WORLD.'

LONDON:
MURRAY, ALBEMARLE STREET.
1859.

The answer is that almost all the babies die before they can breed. For most sorts of living thing, this means the total population stays about the same all the time.

The babies – let's say baby rats – are all different from each other. Some can run faster than others, have a better sense of smell, or have sharper teeth.

They all compete with each other for things like food, shelter and mates.

The 'best' rats find the most food and the safest shelters, so they live long enough to breed. The ones that are slower, or less good at sniffing out food, die.

The rats that breed pass on what made them the 'best' ones (being good at running or biting or smelling) to their babies.

This goes on and on over many years, and the rat population slowly changes into… **super rats!** They're so different from the original rats that they have become a new **species**.

Natural selection

This happened over and over again for millions of years – not just with rats, but with all living things. The world started off with just a few simple animals and plants, but more and more appeared as time went on. Now there are millions of different species.

One question Darwin couldn't answer, however, was *why* the rats were different from each other in the first place?

We know now that it's because of their DNA.

So What is this DNA Stuff Anyway?

Your cells are too small to see without a microscope. Five hundred of them side by side would only measure one centimetre. But you have two metres of DNA in each cell! Altogether you have *thousands* of kilometres of DNA in your body.

It must be pretty important then . . . You need to understand a bit about DNA before the rest of this book will make sense. Here goes . . .

What is DNA?

※ Your body is made up of billions of cells.

※ Every cell contains long threadlike things called **chromosomes**, made of a substance called DNA.

※ DNA is very, very long and thin. It's shaped like a twisted ladder (sometimes this is called a **Double Helix**).

※ The rungs (steps) of the ladder are made of pairs of **molecules** called **bases**.

※ There are four different bases, called A, C, G and T. They fit together in pairs. A fits with T, and C fits with G.

※ The order of the bases down one side of the ladder forms a code – a bit like an alphabet that only has four letters in it.

T (Thymine)

G (Guanine)

What does DNA do?

DNA is a set of instructions for making a living thing. Each instruction is called a **gene**. Each gene controls one thing, like eye colour or hair colour or nose shape. The order of the bases on the DNA makes up each gene. Many genes together make up each chromosome. It seems incredible that something as simple as DNA can make you, or a tree, or a frog.

DNA was discovered in 1869, but it was ages before anyone realised how important it was. A four-letter alphabet seemed much too simple to control everything that happened inside a cell.

The next surprise about DNA is that only about ten per cent of it actually makes up genes. The 'junk DNA', as we call the non-gene stuff, is still very important though. Scientists now think that it helps to switch genes on and off at the right times. We also use junk DNA for **DNA fingerprinting**.

There's more about DNA later in the book.

A (Adenine)

C (Cytosine)

9

The Monk and the Pea Plants

At about the same time as Darwin was working out his ideas about evolution, Gregor Mendel was counting peas...

Mendel (1822-1884) was a monk in a monastery in what is now the Czech Republic. He wanted to be a Biology teacher as well as a monk, but he failed his exams twice, so he had to change his plans and teach maths and physics instead. (Is that really easier?) As well as teaching and doing all the things that monks were meant to do, he was in charge of the monastery gardens. He spent a lot of time doing experiments on pea plants. He was interested in how things like the colour of the flowers or the length of the stems passed from one plant to its offspring.

Here's what Mendel did:

1. He took pollen from the flowers of a tall pea plant and put it into the flowers of a short pea plant.

2. He collected the seeds of the short pea plant and grew them to see if they turned into tall or short plants.

3. He found they were all tall.

4. He took pollen from one of these new tall plants and put it into the flowers of another *new* tall plant.

5. He collected the seeds of that plant and grew them to see if they turned into tall or short plants.

6. This time, there were tall plants and short plants. There were about three times as many tall ones as short ones.

Mendel did this over and over and always got the same results.

Gregor Mendel worked out the laws of inheritance before anyone knew genes existed.

A great discovery?

Mendel told other people about his work in 1865, but no one was very interested. It wasn't until 1900 that three scientists, working separately, re-discovered the same thing and realised how important it was. Without knowing anything about DNA, Mendel had discovered genes.

11

Your Chromosomes and YOU

You're probably a bit like each of your biological parents, but not exactly the same as them. Have you ever wondered why?

You began life when one cell from your mum (an egg) and one cell from your dad (a sperm) joined together. Most cells in a human have 23 *pairs* of chromosomes. Sperm and eggs are special, because they each have 23 *single* chromosomes. This means that when they join, they make a cell with 23 pairs. So you got half of each pair from your mum and the other half from your dad.

You may have noticed you're not just one cell now… That one cell divided in two, then each of those divided over and over. You turned into a ball of cells called an **embryo**, then into a baby. Now you are made of billions of cells.

As it develops in the uterus a baby is called a foetus.

Sperm cell

Egg cell

Fertilised egg

Embryo

Foetus

Baby

But here's the REALLY clever bit.

How it works

Every time a cell is about to divide, it makes copies of all its chromosomes, so the new cells get a complete set.

The two chromosomes that make up a pair contain genes that control the same things. So you get one copy of each gene from your dad, and the other from your mum. But, the two copies might not be exactly the same... Different forms of the same gene are called **alleles**.

Let's look at the gene for eye colour. We'll only look at brown and blue eyes, to keep things simple.

If both alleles say, have brown eyes, then you will have brown eyes.

If both alleles say, have blue eyes, then you will have blue eyes.

But what happens if one allele says, have brown eyes, and the other one says, have blue eyes? Do you end up with one blue eye and one brown eye? Are your eyes blue one day and brown the next? Of course not!

What usually happens is that one allele is stronger than the other, and that's the allele that takes control. In this example, the allele for brown eyes is stronger than the allele for blue eyes.

So if you had one of each, you would end up with brown eyes.

Gene from dad	+	Gene from mum	=	Child's eye colour
Brown allele	+	Brown allele	=	👁
Blue allele	+	Blue allele	=	👁
Brown allele	+	Blue allele	=	👁

The Battle of the Sexes

You now know that you have 23 pairs of chromosomes in each cell. One is a pair of sex chromosomes. These come in two forms: X and Y. Boys have one X and one Y. Girls have two X chromosomes.

When sperm and eggs are made, they only get *one* of each pair of chromosomes.

Every egg gets an X. Sperm either get an X or a Y.

This means that the sex of a baby depends on which type of sperm joins with the egg. There are equal numbers of X sperm and Y sperm, which is why there are equal numbers of girls and boys born.

It's the same in all other mammals, but in birds and butterflies males are XX and females are XY. In some insects, females get two X chromosomes and males get one X and there's no Y at all.

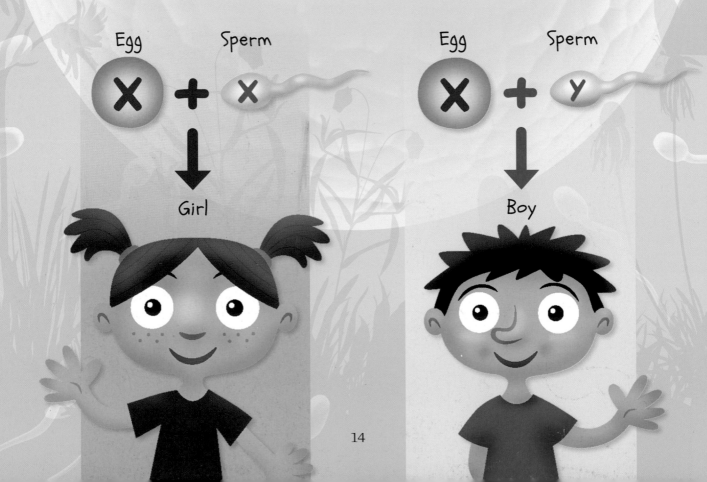

Egg   Sperm   Egg   Sperm

X + X   X + Y

Girl   Boy

Sometimes things can go wrong in humans and people end up with the wrong number of sex chromosomes.

most interested in. It used to be thought that a greater percentage of men in prison had XYY chromosomes than the percentage normally present in the non-prison population. This led some lawyers to argue in court that their clients weren't responsible for what they'd done – they blamed their bad behaviour on the extra Y chromosome they had. It wasn't a very good excuse though, since most XYY men are law-abiding citizens and current research suggests that the results from the previous studies were incorrect.

What can go wrong?

Human embryos can't develop with only a Y chromosome. They can sometimes develop with just one X, or an X and *two* Y's (XYY), but this can cause problems with development. Women who have three X chromosomes (XXX) seem to be normal. XYY is the combination that people have been

I don't know Y I keep stealing things ...

This male chaffinch is XX.

Francis Crick

Maurice Wilkins

James Watson

The Great DNA Race

Once scientists realised how important DNA was, lots of them started trying to find out what it looked like and how it was put together. This was going to be a very important discovery, and it became a race between different groups of scientists to try and work it out first. Whoever won the race was going to be famous.

Francis Crick and James Watson worked at Cambridge University in the United Kingdom. They were supposed to be doing research into other things, but they became very interested in DNA and started to spend all their time on that. At King's College in London, England, Maurice Wilkins and Rosalind Franklin were also trying to find out more about DNA.

You couldn't have found four more different people! James Watson was a young American who had just come to Cambridge. He was very, very clever. In fact, he'd gone to university in America when he was only 15 years old, and he was only 23 when he arrived in Cambridge. He was determined to find out what genes were – preferably without having

Bet you never knew scientists had races!

16

to do too much work. Francis Crick was already working in Cambridge. He was very clever, but hadn't done anything important yet, although he was already 35.

From bombs to DNA

Maurice Wilkins had been working for years in the same laboratory at King's College. During the Second World War he had helped to design the atomic bomb. In 1951 he had been looking at DNA for some time, and it was his work that had made Watson and Crick interested in the subject. British scientist Rosalind Franklin had been working in Paris,

Rosalind Franklin

which she loved. When she came back to the United Kingdom to work at King's College she found life very different. She and Maurice didn't like each other, so it wasn't a very happy laboratory to work in.

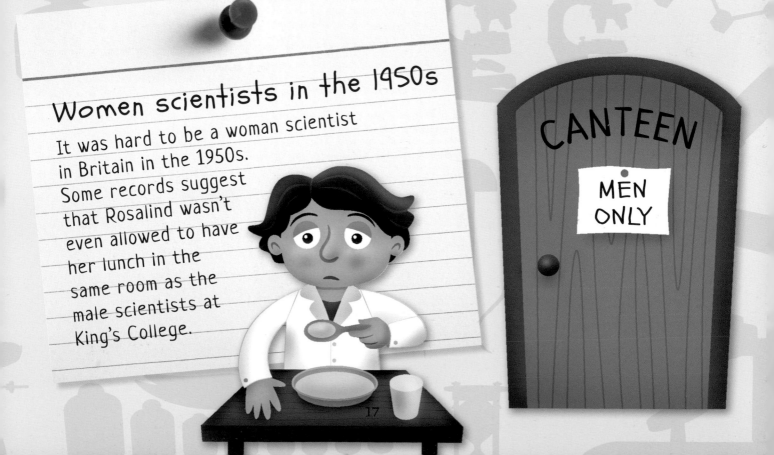

Women scientists in the 1950s

It was hard to be a woman scientist in Britain in the 1950s. Some records suggest that Rosalind wasn't even allowed to have her lunch in the same room as the male scientists at King's College.

CANTEEN

MEN ONLY

Taking Photos of DNA

Maurice Wilkins and Rosalind Franklin both took special photos of crystals of DNA using X-rays at Kings College. These don't look much like normal photographs, so it's very hard to work out exactly what they show. Maurice was good at this, but Rosalind was even better. She was a very serious scientist, and always wanted lots of evidence before she told other people about her results.

This couldn't have been more different to James Watson: all he wanted was to get results as fast as possible. Crick and Watson thought the best way to work out what DNA looked like was to build models of it, and try to make the models match what other people had already discovered about DNA. Crick had already been using models like this in his other work, so it wasn't such a daft idea as it sounds.

The secret 'borrowing' of Rosalind's research

In March 1952, Jim Watson went to see Maurice Wilkins to talk about DNA. While he was there, Maurice showed him one of Rosalind's best DNA photos – without her permission. Watson was able to get a lot of information from looking at it.

On the 7th March 1953, Watson and Crick finished a model that fitted correctly with everything that was known about DNA at the time, including the information they had

Rosalind Franklin's best DNA photo, know as 'Photo 51'.

The structure that Watson and Crick worked out from the 'borrowed' photograph.

Winning the Nobel Prize

Wilkins, Watson and Crick were awarded a Nobel Prize for their work on DNA in 1962. Rosalind Franklin had died of cancer (possibly because of all the work she did with x-rays) in 1958. She was only 37 years old. The Nobel Prize is only given to living people, so she couldn't be given a share in it.

secretly learned from looking at Rosalind's photo. Francis Crick went home and told his wife that he'd found something very important, but she didn't take much notice. He was always saying things like that, and she'd got used to ignoring them. But this time, he was right. However, it was another 25 years before experiments proved that their ideas about DNA were correct. Sometimes, science is very slow!

Everything else in this book is based on these scientists' amazing discovery . . .

The secret of life

When Watson and Crick realised they had won the race to find the structure of DNA, they went to the Eagle pub in Cambridge and told the people there that they had discovered 'the secret of life'.

DNA Double Helix 1953
"The secret of life"
For decades the Eagle was the local pub for scientists from the nearby Cavendish Laboratory.
It was here on February 28th 1953 that Francis Crick and James Watson first announced their discovery of how DNA carries genetic information.
Unveiled by James Watson
25th April 2003

The Human Genome Project

In 1990, scientists all over the world started work on a huge project called the Human Genome Project, to find out the order of all the bases in the DNA of a human. This was a huge task: there were about 3,000,000,000 DNA bases to check! They expected the job to take 15 years, but they finished in 2003, two years early. This was because better computers had made the work much faster.

Now that scientists know the correct order of the DNA bases, they know what each gene should look like. From this, they have worked

Part of a printout from an automatic DNA sequencing machine.

See pages 8-9 for an explanation of DNA bases.

out that mistakes in the DNA order can cause certain diseases. This is because a gene can't give the right instructions if there's a mistake in it.

What does this mean?

The Human Genome Project means we now have tests that can detect some diseases before they make you ill. They do this by looking for a harmful gene. This sounds like a good thing. However, having a harmful gene could also make it harder to get certain jobs or borrow money to buy a house, so it's not all good news.

Owning genes

There have also been some very strange results from the Human Genome Project.

Certain drug companies can 'own' genes if they are able to work out the order of the DNA bases on them before anyone else. This means no one else can use that gene for research without the company's permission for 20 years after it's been discovered.

So if a drug company found that you had some amazing, useful gene and worked out the order of the DNA bases on it, *they* could own it. That means they could make money from it, but you couldn't, which doesn't seem very fair at all.

We're all the same, really!

One thing that the Human Genome Project has shown is just how alike all humans are all over the world. Over 99.9 per cent of your DNA is exactly the same as every other person on the planet! Even more amazingly, you also share 98.5 per cent of your DNA with chimps, and 97.5 per cent with mice!

Even our mother can't tell us apart!

Embryo Selection

Some diseases caused by mistakes in DNA are passed on from parents to their children. If we know what mistake in the DNA to look for, it is possible to test parents to see if they have a chance of passing on these diseases. If there *is* a risk of passing on a genetic illness, the embryo can *also* be tested to see if it has the harmful gene. (An embryo is the tiny ball of cells formed when a new cell created by a sperm and an egg starts to divide and multiply.) This information can be used to help couples who risk passing on a harmful gene to have a baby that will definitely be healthy. This is called embryo selection.

Making babies

Couples who are at risk of passing on some inherited diseases can now have their DNA checked to see if it's faulty. If it is, and they want to have children, they can choose to have what's called in vitro fertilisation. This is where eggs and sperm are mixed in a glass dish in a laboratory – to make test tube babies.

In vitro fertilisation and embryo selection

(in vitro = in glass)

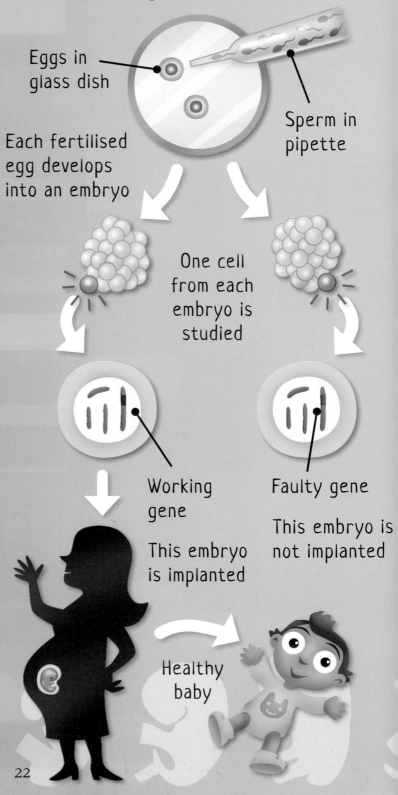

Eggs in glass dish

Sperm in pipette

Each fertilised egg develops into an embryo

One cell from each embryo is studied

Working gene

Faulty gene

This embryo is implanted

This embryo is not implanted

Healthy baby

How it works

When a sperm and an egg fuse together they make an embryo, which starts to divide and multiply: one cell becomes two, two become four, four become eight and so on.

Scientists can remove a single cell from each embryo without damaging it, and check the DNA to see if it's faulty or not.

They can then **implant** an embryo with no genetic faults into the woman's **uterus** to grow into a healthy baby.

As we become able to test for more and more faulty genes, we will be able to prevent more and more babies being born with inherited diseases. Most people think this is a good thing.

A single sperm can be injected into an egg.

> But where do you stop once you start?

The right to choose

Should you be able to choose whether you want to have a boy or a girl? Should you be able to say you don't want a baby who is colour-blind? Or left-handed? Many people worry that we may have too much power to change what our children will be like in the future. Once you start selecting embryos based on which genes they do or don't have, where do you stop?

- ☑ Blue eyes
- ☒ Brown eyes
- ☒ Black hair
- ☑ Red hair
- ☑ Sporty
- ☑ Clever

Gene Therapy

The Human Genome Project has allowed us to identify the faulty genes that cause some very serious illnesses. If scientists can repair or replace these faulty genes, they might be able to cure some of these diseases. This is called gene therapy. We've only recently started to use it, and it isn't easy to do.

It has already cured some people.

How it works

1. First scientists have to make lots of working copies of the gene.

Working gene

2. The next step is to smuggle the working gene into the cells that contain the faulty gene, in the hope that the working gene will take over.

Faulty gene

3. To do this the working gene is hidden inside viruses that have been altered to make them harmless, or inside tiny bubbles of liquid fat.

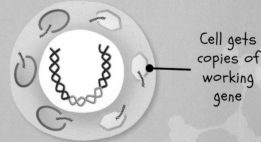

Working gene in fat bubble

Working gene in harmless virus

4. Finally, the virus or fat bubbles are taken in by cells containing the faulty gene.

Cell gets copies of working gene

5. If everything goes to plan, the replacement gene goes to work in the cell, and the effect of the faulty gene disappears.

Scientists have tried to use this technique to treat and even to cure cystic fibrosis. They asked sufferers to breathe in the working genes, as with cystic fibrosis, it is the lungs that are affected. Unfortunately, so far this treatment has not been very successful.

Remember, science can be slooooow!

Replacing genes

Since the 1990s there have been many attempts to use gene therapy. It has been used to help people who suffer from a particular type of inherited blindness to see again, and to treat some forms of cancer. It hasn't yet been the big success that scientists hoped, but it's early days yet.

Genomic Medicine

This is a new type of medicine that's only possible because of the Human Genome Project. It involves trying to make personalised drugs for people, based on their genes.

Some of you reading this will have asthma and will use an inhaler.

One of the most common drugs in these inhalers is called 'salbutamol' in the UK, or 'albuterol' in the US. It works really well for a lot of asthma sufferers. Inhalers containing this drug work by making the muscles in the walls of the airway relax. This makes the airway tubes wider, so it's easier to breathe. However, for some asthma sufferers, 'salbutamol' doesn't work at all. Looking at the genes of people with asthma has explained why.

The reason some asthma sufferers find 'salbutamol' useless is that their muscles contain a slightly different protein than is found normally. This is because the gene that instructed the muscle to grow is slightly different. These people can be given a different drug in their inhalers, which will do the same thing. Doctors don't check people's DNA to find out which drug to give them though – it's much easier to just try them on 'salbutamol' and see if it works or not!

What is a protein?

→ Your muscles, body tissue, bones and organs are made of different proteins. Proteins are made of **amino acids**.

A protein

A chain of amino acids

→ Your cells then use the amino acids to build all the different proteins they need.

Amino acids are put together

→ Your body can make some of these, but to get others you need to eat protein rich food like meat, dairy products, nuts, seeds and pulses.

→ Your DNA tells the cells which amino acids to put together, in which order, to make the right protein.

Amino acids can fit together in different ways

All these foods contain protein

→ Your body breaks down the food during digestion to release the amino acids.

Amino acids are released

It's like using the same building blocks to build different models.

Clones

The idea of 'clones', lots of exact copies of one person or other living thing, is something you often read about in comics or see in science fiction films. Most clones in stories are scary or dangerous, and usually working for the baddie and causing trouble, but in real life clones aren't monsters at all. In fact, some of you reading this are clones.

That surprised you, didn't it? But it's true, honestly.

Non-identical

Two eggs are released

They are fertilised by two sperm

They develop into two embryos

Each embryo develops into a non-identical baby

Identical

One egg is released

It is fertilised by one sperm

It develops into one embryo. The embryo splits in two

Each embryo develops into an identical baby

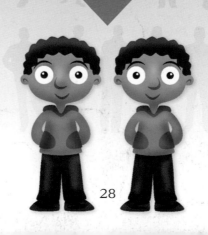

Conjoined

One egg is released

It is fertilised by one sperm

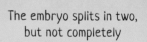

The embryo splits in two, but not completely

Each embryo develops into an identical baby, linked by some part of their bodies

Clones are living things that have identical DNA. All their genes are identical, which makes them… identical. Like identical twins. Or triplets. Although not all twins (or triplets) are clones. There are two types of twins, identical and non-identical. Identical twins are always both boys or both girls and look very similar. Non-identical twins can be both boys, both girls, or one of each sex.

Identical or non identical: how twins happen

Sometimes, two eggs are released at the same time from a woman's ovaries. If each of them is fertilised by a sperm, each one develops into an embryo, and then a baby. The two babies are in the uterus at the same time, but they are no more alike than any other brothers or sisters (or one of each). These are non-identical twins.

Sometimes when a single egg has been fertilised normally by a single sperm, the embryo splits into two when it is still just a tiny ball of cells. Each half develops into a complete, normal baby. But they have the same DNA, so they are identical twins.

Very occasionally this goes wrong, and the embryo splits, but not completely. This is how **conjoined twins** develop. How they are linked depends on which bit of the embryo doesn't split. Conjoined twins used to be called Siamese twins, after the famous twins Chang and Eng Bunker who were born in Siam (which is now called Thailand) in 1811. They came to America as 'curiosities' in a carnival and stayed to become farmers. They married sisters, and Chang had 10 children while Eng had 11! They died on the same day in 1874.

The famous Chang and Eng Bunker.

Dolly the Sheep (1996–2003)

The most famous clone isn't a human; it's a sheep – Dolly the sheep. However, Dolly isn't a clone because she's a twin. She was made in a different way.

How Dolly was cloned

Finn Dorset ewe

Scottish Blackface ewe

Cell taken from udder

Unfertilised egg

Udder cell and 'empty' egg cell made to fuse together by a tiny electric shock

Nucleus removed and destroyed

'Empty' egg with no DNA

Another tiny electric shock starts cell division and an embryo develops

Embryo is implanted into the uterus of another Scottish Blackface ewe

Dolly is born.
She is a clone of the Finn Dorset ewe because all her DNA came from that sheep

Making history

Dolly is famous because she was the first mammal to be cloned like this, though Professor John Gurdon cloned frogs over 50 years ago. Dolly was cloned near Edinburgh, in 1996, by a team of scientists led by Professor Ian Wilmut. He named her Dolly after Dolly Parton, a famous American country and western singer. Dolly had four lambs of her own by a ram called David. They weren't clones, they were just normal lambs.

Dolly got a lung disease that's normally found in much older sheep, and was put down, aged six, in 2003. Sheep usually live to be 11 or 12, so this was very young. Some scientists think that it's because Dolly's DNA had come from a four year old sheep. This means her DNA was already four years old when she was born, so the DNA that made her was 10 years old when she died.

How much do you love your pets?

Since Dolly was made, scientists have cloned cats, dogs, cows and monkeys. Some very rich people have even had clones made of their pet cats and dogs when their pets die! Cloning dead pets doesn't come cheap though. It costs about $150,000 dollars, which is about £93,000 pounds per pet.

No human cloning allowed

No one has managed to clone a human yet – in fact there are rules to stop this happening in many countries and states around the world. This is because many people are concerned that cloning may not be used in an **ethical** way.

Stop! Catch that thief!

DNA Fingerprinting

Remember the 'junk' DNA you learned about on page nine? It's time to find out just how useful it is. We can use it to find out who has committed a specific crime, who is related to who, and whether or not you are being charged too much for food in a restaurant.

The more closely related you are to someone else, the more alike your junk DNA will be. You can take the junk DNA out of the cells in a blood sample and chop it up. If you put the chopped DNA in a hole in some special jelly and pass electricity through the jelly, the bits move apart to give a pattern of stripes. The more alike the

DNA pattern.

patterns of two different people, the more closely related they are.

This technique is often used to help solve crimes now. A sample of DNA can be taken from, say, a bloodstain. The DNA in the strain can be compared with DNA samples from suspects to look for a matching pattern.

Solving crime

DNA fingerprinting was first used to solve a murder in 1988. It proved that a teenage boy who had been charged with a double murder must be innocent, because his DNA didn't match the samples from the bodies. It also proved that another man must be the murderer because his DNA *did* match the sample. He was sent to jail for 30 years.

Cold cases

DNA fingerprinting can even be used on bloodstains that are years old, so some crimes from long ago have recently been solved using new DNA evidence. It has also been used to prove that some people who were jailed in the past for various crimes were actually innocent. More than 160 people have now been freed after DNA fingerprinting showed they hadn't carried out the crimes for which they were in prison.

I'm free!

C'est bon!

BUDGET CAVIAR

Caviar is a dish made of the raw eggs from a fish called a sturgeon.

But what about that restaurant bill?

Some very expensive restaurants sell something called caviar. Traditionally, caviar is a dish made of the raw eggs from a fish called a sturgeon. DNA fingerprinting of the fish eggs shows that some restaurants mix the sturgeon eggs with much cheaper fish eggs, so they make more money.

Sturgeon fish.

DNA and Forensic Science

Many people love reading crime novels or watching crime shows on TV. But do these shows resemble reality at all? What can the police and forensic scientists really work out using only DNA? And is DNA evidence always reliable?

DNA fingerprinting has been used as evidence in hundreds of cases since 1986, from murders and serious violent crimes to break-ins and burglaries. As DNA fingerprinting can now be used on bloodstains that are many years old, sometimes cases that seemed to have gone 'cold' in the past can now be solved and the criminals brought to justice.

So, yes, DNA really is used to solve crimes.

But on TV the forensic scientists get results in a couple of hours, whereas in real life it can take weeks…

The UK National Criminal Intelligence DNA Database

Britain's database of DNA samples taken from anyone arrested and detained at a police station was created in 2005. It now holds more than 3.1 million DNA records taken from both police suspects and crime scenes. Whenever a new sample is added to the database, an automatic search takes place that flags up any links between samples. This can be used to link crime scenes to potential suspects.

Dusting for fingerprints.

34

Finding DNA evidence

Fingerprints.

Fibres from clothing.

Hair.

Blood.

Hopefully, you'd have an **alibi**...

Where do we find DNA?

DNA can now be picked up from a single hair or a few skin cells, and copied in a machine until there's enough for DNA fingerprinting. This is great because even the tiniest trace of evidence is useful, but it also means that the people collecting the samples have to be terribly careful they don't **contaminate** the crime scene.

Contamination really is a serious problem. If you gave someone a hug, and then they robbed someone's house, still wearing the clothes they had on when you hugged them, your DNA would be all over the crime scene!

Mice with Four Parents

What? How can that make sense?

Sometimes two embryos join together at a very early stage of development to make one baby. In a way it's the opposite of how you get identical twins.

Scientists can mix embryos together in the laboratory, to create animals that have four parents. They can join two mouse embryos, one from parents with black hair, and one from parents with white hair. This makes a mouse with a mixture of black and white hair. This is useful, because it tells the scientists something about what bits of which embryo turn into what bits of the adult.

What's interesting is that the new mouse also has two kinds of DNA. One kind from each original embryo. This is called being a **chimera**, and amazingly, occasionally it happens naturally in humans too.

1. Two sets of parent mice are chosen

2. Two mouse embryos are joined together

3. A mouse foetus forms

4. A mouse with both white and black hair is born

Human chimeras

Sometimes in humans, when there are twin embryos, they join together very early on and end up as one person. The person probably wouldn't ever even know that it had happened. You wouldn't be able to tell anything was different about that person just by looking at them. The secret would be in their DNA.

The story of Lydia Fairchild

Lydia Fairchild lived in America. She had two children and was pregnant with a third, when her marriage broke up. She was asked to take a DNA test to help prove that her husband was the children's father. The results showed that he was the father, but she wasn't the mother!

Police thought maybe she had stolen the children, and they were taken away from her. When her third baby was born, its DNA was tested right away, and the result also said she definitely couldn't be its mum. Everyone realised that this was impossible, since they'd seen her give birth, but no one could explain what had happened.

When they did more tests, they found that DNA from some parts of her body matched her children's DNA, but DNA from other parts didn't. She had two different lots of DNA, because she was a chimera, non-identical twins who had ended up as a single person.

Wow!

1. Two eggs are released

2. They are fertilised by two sperm

3. They develop into two embryos

4. The two embryos fuse together to make a single embryo

5. The embryo develops into one baby — but with two sets of DNA: one full set from each original embryo

Genetic Engineering

The subject of genetic engineering is often in the news, but what does it really mean? Genetic engineering means taking a gene from one type of plant or animal, and putting it into a totally different one.

It sounds as if doing this could create some very strange animals and plants, but in real life genetic engineering doesn't result in new animals that are half-cat, half-fish, or plants that grow carrots underground and apples on their branches.

Genetic changes

The first food to be changed by genetic engineering was a type of tomato that was given an extra gene to slow down the rotting process. It meant it could be picked when it was ripe (most tomatoes are picked when they are still green) yet stay fresh for a long time in the shops.

However, the company that invented these genetically engineered tomatoes stopped growing them after just a few years. This was because they hadn't chosen a very good type of tomato to work with, so even though their tomatoes stayed fresh for longer, they didn't taste as good as other types of non-genetically engineered tomatoes.

From science lab to supermarket

In British supermarkets today, there are at least eight products on sale made using genetically modified plants. These include various types of cooking oil, crackers, and chocolate. They all have to be labelled to say that they are made with genetically modified ingredients. However, milk, eggs and meat made from animals that have been fed genetically modified feed do not have to be labelled.

Genetically engineered crops

ANIMAL FEED

Hungry cows eat the genetically modified feed

The cows are milked

The milk is made into cheese

EASY CHEESE

INGREDIENTS:
Milk, Rennet, Salt, Stabilizers and Preservatives

Feeding the World

One of the reasons scientists are so interested in making genetically modified plants, is that the population of the world is getting bigger and bigger, so there are more and more people to feed.

Using genetic engineering, plants like rice and wheat can be genetically modified so that they become **resistant** to insects and other pests; can survive **droughts** or floods, or contain extra vitamins. This means fewer crops will fail, and more people can be fed using the same amount of land.

Some people believe that genetically engineering crops is the only way to make sure there will be enough food for everyone in the future.

How many?

10

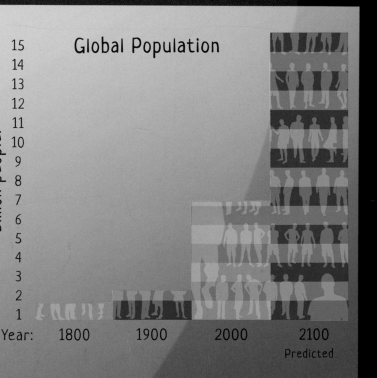

Billion people:

15	Global Population
14	
13	
12	
11	
10	
9	
8	
7	
6	
5	
4	
3	
2	
1	

Year: 1800   1900   2000   2100
                               Predicted

Super weeds and other problems

Other people are worried that if we start growing lots of genetically modified crops, we may accidentally cause much bigger problems in future.

Adding genes to plants could have unexpected **side effects** on the animals or people that eat the plants. The modified plants could **breed** with weeds creating 'super weeds' which would be very hard to kill. And, according to Darwin's theory of evolution, any insects or pests that manage to survive on plants that are supposed to kill them would eventually become 'super pests'.

This is why very few genetically modified crops are grown in Europe at the moment, while scientists work out all the possible risks and benefits.

Are you calling me a weed?

Modifying mosquitoes

Scientists in the UK are experimenting with a new way of reducing the number of mosquitoes that carry **dengue fever**. They have genetically engineered male mosquitoes to produce offspring that will die before they grow into adult, disease-carrying mosquitoes. If it proves to be safe and effective, the company hope the same method could be used on the four species of mosquito that spread malaria, a disease which kills over half a million people every year.

Making Medicine

Advances in genetic engineering have allowed scientists to use bacteria to make chemicals needed by the human body to work properly. For example, we can now make insulin, which is used to treat the illness diabetes.

People who suffer from type 1 diabetes cannot produce insulin on their own. In the past, insulin was taken from the pancreases of slaughtered pigs, cows and sheep. Today, using genetic engineering, we can grow bacteria which produce human insulin instead.

To do this, the human gene for insulin is put into a special kind of bacteria, which is then grown in huge tanks. The bacteria produce the chemical insulin, which scientists can then purify for use in humans. Lots of other chemicals can be made like this too, including human growth hormone, which is used to help children who aren't growing properly.

How insulin is made

The gene for insulin is cut out of human DNA

The gene is inserted into a bacterium

The bacterium is put in a tank where it grows into many bacteria — all of them containing the human gene for insulin

The insulin is extracted and purified, and is then ready to use by humans

I wonder how tall it could make me?

Another amazing sheep

It's not just bacteria that can be modified in this way, but animals too. In 1990, scientists genetically engineered a sheep called Tracy, who had an added human gene as well as all her sheep genes. This let her make a special **protein** in her milk which is normally only produced by humans. This protein could be taken out of her milk and used by doctors to treat the disease cystic fibrosis.

Six years later the same scientists made Dolly the sheep. When they made Dolly they were trying to work out how to clone a sheep like Tracy with an extra human gene safely.

Scientists have also created glowing mice, rats, cats, pigs, and even monkeys, and are hoping that studying these animals will help them to understand more about how moving genes between species might help to fight some human diseases.

Proteins are the building blocks that make bones, tissues and organs.

Glow in the dark pets!

Today, you can even buy genetically engineered pets! GloFish, for example, are goldfish that have been genetically modified by adding the gene that makes some jellyfish glow in the dark. They come in red, orange, blue, purple or green.

It's very hard to get to sleep with the lights on!

Mice with Human Ears

Pardon?

If you look up 'Vacanti mouse' on the Internet, there's a famous picture of a very strange looking mouse. It doesn't have human ears on its head. It just has normal mouse ears. However, it *does* have what looks like a human ear on its back. Except it isn't actually a human ear at all. Confused? Let me explain...

The thing on the mouse's back is the shape of a human ear, but it isn't made of human cells. It isn't made of mouse cells either. It's made of cow **cartilage** cells. Cartilage is the bendy stuff in your ears and at the end of your nose. Scientists put an ear shaped mould under the mouse's skin, then injected cow cartilage cells into the mould. The cells grew over the mould to create the final ear shape. It doesn't work as an actual ear though; the mouse can't hear anything with it.

But why?

Well, human ears often get damaged because they stick out, and they're very hard to repair because they're such a complicated shape. Plastic surgeons had often said that they would like to have 'spare ears' to transplant onto people who had lost or damaged their own, and so scientists began to experiment to see if this might be possible.

Could we grow spare ears on humans?

We could, but it wouldn't work with cow cells, because your **immune system** would attack the new ears as though they were a dangerous infection, and they would probably go black and fall off. (Your immune system is what fights off germs. It recognises and destroys cells that shouldn't be in you. That's good for stopping you getting ill, but it also means it will destroy organs transplanted from another person.)

However, scientists can now produce new ears using human cartilage cells – grown on a special ear 'scaffold' in a dish, not on a mouse – and are beginning to transplant them onto children who are born with abnormal or missing ears.

Because they do this using the children's own cells, the ears shouldn't be rejected by the children's immune systems. The same technique is also being used to grow new noses and even replacement **arteries**.

A replacement ear being grown.

What are Stem Cells?

On the news, you might often hear the words 'stem cells' being mentioned, and lots of talk about what scientists might be able to do with them one day. But what are they? Well, they're nothing to do with plants...

Stem cells are cells that haven't decided what they're going to be yet, and can still turn into any type of cell. They can also divide many times.

This makes them very exciting for scientists.

How are cells different?

All living things start out as just one single cell. Most living things end up as billions of cells all working together. To get from one single cell to the final fully-grown human being or cat or potato or rose bush, the following things need to happen:

1. The cells need to multiply. First, the single cell divides into two cells.

2. Then the two cells grow until they are fully sized, then divide again, resulting in four cells, then eight and so on.

3. This happens over and over again.

4. The cells must become different from each other.

5. If they didn't do this, you would just be a giant blob, not a person with skin and muscles and bones.

6. Most living things, including *you*, are made of lots of different types of cells, each of which does a different job.

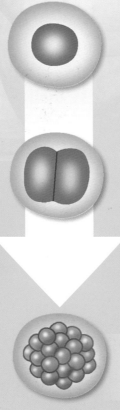

Genetic information

All of your cells have a full set of genes in them, but most of them only use a few of those genes. For instance, white blood cells only use the genes that make them good at fighting diseases. They don't use the same genes as skin cells, even though both types of cells have all the genes.

Once a cell has 'decided' what it's going to be, the genes it won't use are switched off, and can't usually be switched on again. Usually this also means the cells won't divide very much any more.

But unlike all these other cells with particular jobs, stem cells can still turn into *anything*, and they will also keep dividing.

This is what makes them so exciting!

A single fertilised egg has the genes to make the whole organism.

Growing New Organs

Scientists want to grow stem cells in laboratories in order to work out how to control the genes that make them turn into different sorts of cell. If they can get this to work, it might mean that in future they could grow new organs like kidneys and hearts, which could be used for transplants when people become ill.

Organ transplants

Every day in the UK, about three people die while waiting for an organ transplant. The people who are lucky enough to receive a **donor** organ also have to take drugs for the rest of their lives to stop their immune systems from attacking the transplant. If new organs could be grown from a patient's own cells, patients wouldn't need to wait for donors or to take drugs for the rest of their lives.

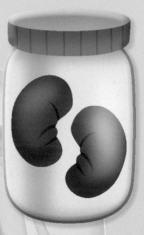

Hearts

Kidneys

Intestines

Will there be one if I need it?

Saving lives

In 2008, something like this was done for the first time, when scientists grew a windpipe (the tube that takes air to your lungs) using a patient's own cells and used it to replace a damaged one in a woman who was very ill.

In this case they took a donor windpipe and used chemicals to get rid of all the windpipe cells, leaving only the cartilage behind. They then used stem cells taken from the sick patient to grow new windpipe cells over the cartilage. The operation was a success, and the patient was cured and also did not need to take any drugs.

Healing hearts

Doctors have also injected stem cells into damaged hearts to help to repair them after heart attacks. Research is still being carried out but doctors hope that one day they will be able to cure **heart failure** this way.

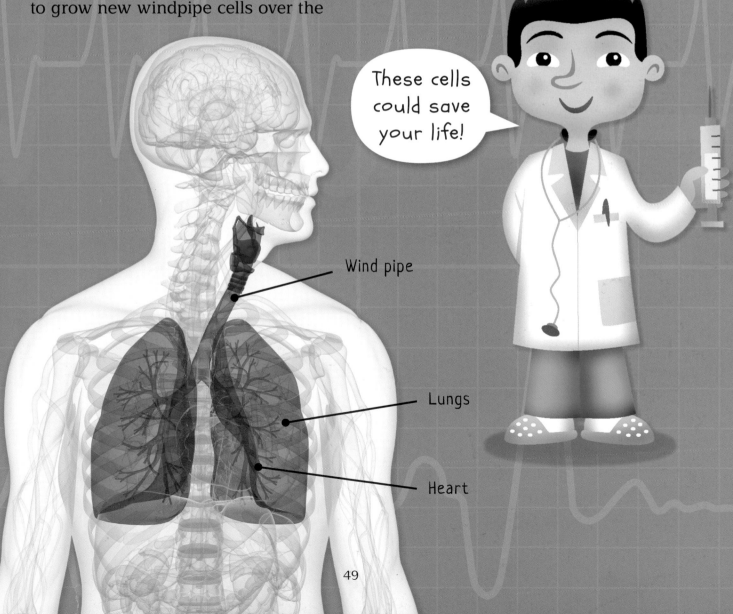

Wind pipe

Lungs

Heart

Where do Stem Cells Come From?

It sounds as if stem cells are a great idea, doesn't it? Then why do some people get so upset about using them? This is because of where stem cells come from.

Embryo stem cells

The idea of taking stem cells from embryos upsets some people. They say that as soon as a sperm and egg join, what you have is a human being, even though it's still just a single cell. Other people think that a ball of cells is just a ball of cells and not a human, so it is acceptable to use embryos to collect stem cells. Everyone is allowed to have their own opinion about whether this is right or wrong.

Stem cells can come from:

1. Embryos. When a sperm and an egg join they make a new single cell. An embryo is the tiny ball of stem cells made when the new cell starts to divide and multiply.

2. Blood from a baby's **umbilical cord**.

3. Some types of adult cells, like bone marrow, which is the fatty stuff inside your bones that makes your blood.

Adult stem cells

Using cells from adults is much more difficult. This is because some of the genes have already been switched off. The trick is to find a way to switch them all on again, so the cells can turn into any other type of cell. If you could do this, you could take cells from your skin and grow them into new bone, or heart, or liver…

Umbilical cord stem cells

The umbilical cord is what joins a baby to its mum when it's in her **uterus**. The baby gets food and oxygen through it until it's born. Once the baby is born, it doesn't need the cord any more, and it's cut off. The scar where it joins the baby to its mum turns into its tummy button.

The cord usually just gets thrown away, but it's got lots of blood cells in it, and lots of stem cells too. Most people are not upset by this way of collecting stem cells.

You have to come out sometime you know!

Umbilical cord

The baby doesn't need its umbilical cord anymore.

51

Saviour Siblings

Fanconi anaemia is a rare and very serious disease that is sometimes passed on from parents to their children. Until recently there was no effective treatment for most sufferers, and many died when they were children.

The Nash family's daughter, Molly, was born with the disease. In 2000, when Molly was six, doctors in Chicago helped her parents to **conceive** the world's first **saviour sibling**. Sperm and eggs from her parents were mixed in a laboratory and allowed to develop into tiny embryos. These were tested to find one that *didn't* have the disease, and had exactly the right cells to help make Molly better. This embryo was put back into Molly's mum's uterus and grew into a perfectly normal baby. His name is Adam Nash.

Baby to the rescue

When Adam was born, his umbilical cord was cut. But instead of throwing it away, stem cells were collected from it and used to replace Molly's bone marrow, which wasn't working properly. Adam's stem cells cured Molly.

Saviour siblings

1. An embryo that is a suitable DNA match is chosen, which grows into baby Adam.

2. When he is born, Adam's umbilical cord is cut off.

3. Stem cells are collected from the blood in the cord.

4. They are grown in a laboratory until there are lots more stem cells.

STEM CELLS

5. Adam's stem cells are used to replace Molly's bone marrow.

Molly is cured!

This treatment was, and still is, controversial (it divides people's opinions) for three reasons:

❋ Some news articles about this kind of treatment are so badly written that it makes it sound as though the 'saviour' baby is hurt in some way in order to treat its sibling, which isn't true.

❋ Some people were worried that children born in order to save their siblings will feel unloved, and think they were only born because they were needed for their stem cells.

❋ In order to make the 'perfect' embryo, there are usually other embryos that don't get used and are destroyed instead. Some people see this as destroying a human life.

Everyone is allowed to have their own opinion about how they feel about this kind of treatment. If it were you, how would you feel if you knew you'd saved your big brother or sister's life, just by being born?

What's Next?

You are living at a very exciting time for biology and medicine ... In your lifetime, it looks as if doctors will be able to use what we are learning about DNA to cure all sorts of diseases that we can't treat now by using gene therapy and genetic engineering. We may be able to use stem cells to grow new organs and repair damaged spinal cords to help disabled people walk again.

Of course, all these wonderful techniques can be misused too. Do we really want people to be able to clone themselves, or to choose everything – even the eye colour – of the baby they're going to have? Is it a good idea for the DNA of all sorts of plants and animals to be mixed together?

The future's bright

We can't do any of these things yet, but we might be able to, one day. This means it's important that there are rules about what scientists are allowed to do. It also means it's important that the people who will make the rules – you, for instance – understand what's going on. I hope this book has made you want to find out more.

Just think, some of you might turn out to be the scientists who make another big discovery about DNA and genetics, and who add another chapter to this amazing life story. After all ...

... now you know what makes you, YOU, what's stopping you?

Things to do: Getting DNA Out of Fruit

Have a go – it's great fun! But you must get a grown-up to help you, as this experiment involves hot water and a chemical called methylated spirits, which should only be used under adult supervision.

What you need:

- **Fruit:** 50g hulled strawberries *or* half a peeled kiwi fruit *or* half a peeled banana

- **Extraction fluid:** make this by mixing 100ml water, with 10ml washing up liquid and 3g (about half a teaspoon) of salt

- **Coffee filter paper**

- **Sieve**

- **Two large empty bowls**

- **One bowl of water at about 60 degrees Celsius:** two parts of boiling water to one part of cold water is close enough if you don't have a thermometer. **Make sure you get an adult to help.**

- **Self-seal plastic bag**

- **Ice**

- **Fresh pineapple juice:** about half a teaspoon

- **A clean, tall and narrow glass**

- **Ice-cold methylated spirits:** only to be used under adult supervision. Methylated spirits can only be bought from a chemist by an adult. This should be put in the freezer in a *plastic* bottle. A glass one could explode!

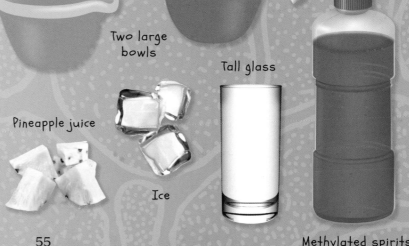

Extraction fluid

Fruit

Coffee filter paper

Sieve

Plastic bag

Bowl of hot water

Two large bowls

Tall glass

Pineapple juice

Ice

Methylated spirits

What to do:

1. Put the fruit (either strawberries or kiwi or banana) into the plastic bag and seal it.

2. Squash the fruit inside the bag until it is mushy.

Squash the fruit

3. Add 50ml of the extraction fluid you have made to the bag, seal it again, and give it a good squish to mix it in with the fruit.

Add the extraction fuid

4. Put the sealed bag in a bowl of hot water (which is about 60 degrees Celsius) for 15 minutes.

Heat the fruit and fluid

5. Put some ice and cold water into one of the empty bowls while you are waiting.

Get the icy water ready

6. After 15 minutes, take the plastic bag out of the bowl of hot water and put it in the bowl of icy water to cool down.

Cool the fruit and fluid

7. Put the coffee filter in the sieve and put the sieve over the remaining clean and empty bowl.

Get the filter ready

56

8. When cool, pour the contents of the sealed bag through the coffee filter.

Filter the fruit mixture

9. Pour the liquid that's gone through the filter from the bowl into a tall, narrow glass. You only need about 10ml.

Pour the liquid into the glass

10. Add the pineapple juice and mix. Leave for two minutes.

Add pinapple juice

11. With an adult's help, get the methylated spirits from the freezer and very slowly pour it onto the fruit liquid in the glass.

Slowly add chilled methylated spirits

12. It will form a purple layer on top of the fruit layer.

You might be surprised how much DNA you get!

Whitish strands of DNA should appear, rising from the fruit layer into the alcohol.

Timeline

1665 Robert Hooke is the first person to see cells under a microscope.

1831 Robert Browne discovers the cell **nucleus** (the 'brain' of the cell).

1843 Karl Wilhelm von Nageli discovers the 'ideoplasma' – what we now know as chromosomes.

1856 Gregor Mendel publishes the results of his pea experiments. No one pays attention or realises the importance of what he has discovered.

1857 Darwin publishes *On the Origin of Species by Natural Selection.*

1869 Friedrich Meischer isolates something he calls 'nuclein' from white blood cells in pus. This will later be re-named DNA.

1900 Mendel's results are rediscovered by three other scientists, who realise how important they are.

1952 Robert Briggs and **Thomas King** clone a leopard frog by taking the nucleus from an embryo cell and putting it into an egg cell from which they have removed the nucleus.

1953 James Watson and **Francis Crick** publish a paper in the science journal *Nature*, describing the structure of DNA. Rosalind Franklin and Maurice Wilkins also publish a paper in *Nature*, describing their own results on DNA.

1958 Rosalind Franklin dies of cancer, possibly caused by the high doses of X-rays she used in her work.

1958 John Gurdon clones a frog by transferring a nucleus from an adult frog cell to an egg cell from which he had removed the nucleus. This shows that all the genes must still be present in adult cells.

1962 James Watson, Francis Crick and **Maurice Wilkins** are awarded the Nobel Prize for their work on DNA. Rosalind Franklin is not, because it is only ever given to living people.

1984 Alec Jeffreys develops DNA fingerprinting.

1990 Ashanti DeSilva becomes the first person to be treated by gene therapy.

1996 Dolly the sheep is born, cloned by Professor Ian Wilmut at the Roslin Institute in Scotland.

1996 Dolly the sheep dies.

2000 Adam Nash, the world's first 'saviour sibling' is born.

2008 Claudia Castillo receives a new windpipe, grown from her own stem cells.

2012 The first use of stem cells from embryos takes place, in a trial to treat a form of sight loss called Stargardt's Disease.

2012 The first genomic medicine to treat Cystic Fibrosis is developed.

2012 John Gurdon and **Shinya Yamanaka** are awarded the Nobel Prize for research which led to stem cell therapy and the cloning of Dolly the sheep.

Find Out More

Read

Hox by Annemarie Allen (Floris Books, 2007)

A science-fiction thriller that touches on genetic engineering and animal rights.

Pig Heart Boy by Malorie Blackman (Corgi Children's, 2004)

What's it like to need a heart transplant? What if you were offered a pig's heart?

Internet-Linked Introduction to Genes and DNA by Anna Claybourne, Stephen Moncrieff and Felicity Brook (Usborne, 2003)

An easy to digest introduction to genes and DNA.

Watch

Jurassic Park (Universal Pictures, 1993)

This has just enough plausible science in it to avoid being totally crazy. And who can resist those velociraptors?

Life Story (BBC, 1987)

A wonderful drama based on the race between Wilkins, Watson, Franklin and Crick to unravel the structure of DNA. You will have to play detective to hunt this down on the Internet, as it was never released as a DVD.

Charles Darwin and the Tree of Life (BBC, 2009)

Marvellous documentary on Darwin's *Theory of Evolution*, presented by the equally marvellous David Attenborough.

Visit

The Science Museum in London to see the original model of DNA.

www.sciencemuseum.org.uk

The Wellcome Trust in London to see a printed version of the entire human genome.

www.wellcome.ac.uk

Down House in Kent. This was the home of the Darwin family, where Charles Darwin wrote *On The Origin Of Species*.

www.english-heritage.org.uk/daysout/ properties/home-of-charles-darwin-down- house

The National Museum of Scotland in Edinburgh, to see Dolly the Sheep.

www.nms.ac.uk/our_museums/national_ museum.aspx

Log on to:

http://learn.genetics.utah.edu

This is a great interactive site that allows you to do all sorts of DNA related things in a virtual lab. Try your hand at DNA fingerprinting, or clone a mouse!

www.dnai.org

A good interactive site with lots of information about DNA

www.nobelprize.org/educational/ medicine/dna_double_helix

Try your hand at making a copy of a chromosome!

Glossary

Alibi Reason why you could not have possibly committed a crime

Allele A different form of the same **gene**

Amino acids Chemical building blocks used by your body to make **proteins**

Arteries Muscular tubes that lead blood away from the heart and to the body

Bacteria Micro-organisms (living things) made of just one cell

Bases The chemicals in DNA that make up the four-letter alphabet of the genetic code. The letters are A, C, G and T.

Biological parents The two people who gave you your DNA

Breed Reproduce

Cartilage Tough, bendy tissue that makes up your nose, throat and ears

Chimera Single animal or person produced when two embryos merge at a very early stage of development

Chromosomes Long strands of DNA found in the nucleus of cells

Clone Two or more living things with exactly the same DNA, such as identical twins

Conceive To become pregnant

Conjoined twins Very rare form of identical twins where the twins are physically joined at some part of the body

Contaminate Spread DNA or other substances in a place where they shouldn't be

Controversial Divides people's opinions

Cystic fibrosis Serious lung disease caused by a fault in a person's DNA

Diabetes Illness caused when a person cannot control their **insulin** levels

Digestion Breaking down food in the body into a form that can be used or removed

DNA Deoxyribonucleic acid: the chemical that makes up genes and chromosomes

DNA fingerprinting Way of comparing DNA from different people to find out if they are related. Can also be used to compare DNA from crime scenes with DNA from suspects.

Donor Somebody who gives away a body part, an organ, tissue or blood for the treatment of other people

Droughts Long periods with not enough rain

Drugs Substance used to treat, prevent or diagnose a disease and to lessen pain

Embryo Tiny ball of cells that develops into a new organism

Embryo selection Checking the chromosomes of an embryo to find one that is free from a genetic disease

Evolution Process by which all living things on the planet developed over huge lengths of time from a few, much simpler species

Ethical Ethics are the **morals** that guide our choices. If something is **ethical** it is morally acceptable.

Forensic scientist Someone who studies the evidence left at a crime scene

Fanconi anaemia Disease that causes many kinds of cancer

Gene Length of DNA that controls a single characteristic such as the colour of your eyes

Gene therapy The process of replacing a missing or faulty gene to treat or cure a disease

Genetic engineering Process of inserting a gene from one species into a completely different species

Heart failure Result of damage to the heart, meaning it stops working completely

Human Genome Project International project to work out the entire genetic code of a human being

Human growth hormone Chemical that tells our bodies when and how much to grow

Immune system Body's system for fighting disease, which can also cause rejection of transplanted organs

Implant Embed or place something in the body

Inherited Characteristics received as a result of genes being passed from parent to offspring

Insulin Hormone that controls the body's blood-sugar levels

In vitro fertilisation Mixing eggs and sperm in the laboratory to produce embryos

Nucleus The 'brain' of a cell, where all the genetic material is stored

Molecule Smallest physical unit of a substance

Morals Ideas about what is right and wrong

Pancreas Organ that creates hormones including **insulin**

Population All of the people that live in a particular area

Proteins Building blocks that make body tissue, bones and organs

Purify Remove anything harmful or unwanted

Resistant Unharmed by the damaging effects of something

Sex chromosomes Pair of chromosomes (XX or XY) which determine the sex of a baby

Saviour sibling Brother or sister chosen because they are genetically compatible with an ill older sibling and may be able to help them by donating **stem cells**

Scaffold Temporary supporting framework

Side effects Unexpected and usually undesirable results of an experiment or treatment

Slaughter Kill an animal

Species Group of animals or plants that share lots of characteristics and can breed to produce fertile offspring

Stem cells Cells that haven't yet decided what to be when they grow up. They can turn into any type of cell in the organism.

Test tube baby Baby produced by in vitro fertilisation

Umbilical cord Cord through which a baby in the uterus gets its food and oxygen from its mother. Usually thrown away after birth, but contains lots of stem cells.

Uterus Womb

Index